Revolutionizing Robotics
Female Innovators Disrupting the Field

Jessica Ruby Thompson

Table of Contents

You have to learn the rules of the game. And then you have to play better than anyone else.

— Albert Einstein

Chapter 1. Introduction

In an ever-evolving technological era, robotics is undergoing a significant transformation, shattering conventional boundaries and reshaping our vision of the future. This Special Report, "Revolutionizing Robotics: Female Innovators Disrupting the Field," celebrates the astonishing achievements of women at the forefront of these advancements. Rather than delving into intimidating jargon, we will gently navigate the intricacies of this field, exploring how these ingenious female innovators are leading the charge in reshaping our world. Read on, and be stimulated not only by the elegance of these technological breakthroughs but also by the inspiring and groundbreaking journeys of these trailblazing women. This is a journey that promises both enlightenment and inspiration. Dive in and discover your motivation to invest in this report – because the future of robotics, steered by the feminine touch, is indeed a promising and exciting prospect.

Chapter 2. Silhouettes of Innovation: A Historical Perspective

Delving into the historic perspective of robotics innovation, one inevitably perceives a vivid tableau punctuated by notable silhouettes of women, casting a long-lasting impact on the course of innovation. This chapter weaves a rich tapestry of women's momentous contributions in robotics, shining a spotlight on their indelible mark on this captivating field.

2.1. The Dawn of Robotics: A Retrospective View

Much like the dawn of humanity, the advent of robotics was marked by a blend of curiosity, creativity, and courage. Singled out from the multitude of contributors are women—curious minds coupled with valiant spirits who dared to tread the path less traveled. The inception of robotics in the mid-20th century saw remarkable innovators like Cynthia Breazeal from MIT and Helen Greiner, co-founder of iRobot, pioneering the robotic industry with their invaluable contributions.

Amidst an era fraught with gender biases and societal norms, breaking the proverbial 'glass ceiling' was highly challenging. However, these female innovators embarked upon the journey, intent on shaping the world and leaving an enduring legacy.

2.2. Propelling Paradigm Shift: Female Innovators in the First Wave

The era bubbling with the excitement of the first wave of robotics during the late 20th century saw female innovators like Cynthia Solomon, who co-developed the Logo programming language aimed at children. Alongside were shapers like Radia Perlman, the 'Mother of the Internet,' known for her innovations in networking and network security protocols, playing a pivotal role in enabling the communication between robots and the world.

Interestingly, the intersection between robotics and programming is where female innovators shone significantly. Their proficiency in both realms helped them perceive the broader picture, harnessing their power to innovate and inspire.

2.3. Transitioning to Modern Robotics: The Second Wave

Transitioning into the 21st century signified a dynamic shift in the zeitgeist of robotics. Female innovators like Ayanna Howard, leading authority in the field of robotics, and Cynthia Breazeal, popularly known for her work on social robotics and human-robot interaction, emerged as icons.

It was the time when robotics started being integrated into everyday life - the 'Second Wave' of robotics. Their unique approach, combined with a profound understanding of the applications of robots in society, made these women instrumental in guiding the second wave of robotics.

2.4. Disrupting Norms: Women Triumphing in the Robotics Renaissance

Stepping into the latest Robotics Renaissance, we've seen heightened advancements in AI-driven robotics, led significantly by women. For instance, Fei-Fei Li's contributions to computer vision have revolutionized perception-based tasks in machines, making them more adept at recognizing and deciphering visual data.

Moreover, women have ventured into the business aspect of robotics as well with Amy Frampton, Francisco and Kass Dawson heading Google's Robotics as a Service (RaaS), substantially influencing the commercial course of robotics.

In these contemporary times, women have not just participated in the evolution of robotics as innovators but also as educators, entrepreneurs, and policy influencers.

We thus unfold a compelling panorama that etches the tenacious journey of women in robotics, from its genesis to the present. The future rightly promises an exciting prospect steered by the indomitable spirit of women innovators. This historical perspective, replete with 'Silhouettes of Innovation,' serves as a reminder of the resilient women who transformed the treacherous path of gender disparity into a pathway leading to revolutionary breakthroughs in robotics. Their journey sheds light on the importance of diversity in invention, underscoring the essence of having multi-faceted perspectives fostering groundbreaking developments in technology. In these women, we find the potent catalysts that continue to drive the evolution of this fascinating field.

Chapter 3. Discovering the She-Force: Profiles of Leading Female Innovators

In the swiftly changing landscape of robotics, there are women who stand as pillars, guiding the blossoming field towards inclusivity and innovation. These women step into the arena not only to challenge the status quo but also to redefine it, proving beyond doubt that technology is undeniably female as much as it is male. Turning to these stunning profiles that outline the lives, inspirations, and achievements of these female trailblazers, we desire to expand your perspective and encourage both current and future innovators to follow the beacon of their insights and triumphs.

3.1. The Reigning Robotics Queen: Cynthia Breazeal

Cynthia Breazeal holds an uncontested spot among the leading female innovators in robotics. Her inspirations trace back to 'Star Wars,' where the movie's sociable and helpful robots captivated her imagination. Admiring this prototype of the future, she decided to translate this fiction into reality, culminating in the creation of Kismet, the world's first sociable humanoid robot at MIT's Media Lab.

Kismet opened avenues for what Breazeal termed "social robotics," cornering a new space where robots could exhibit emotions to increase user-friendliness and render technology accessible to all, including the uninitiated. As the founder and current Chief Scientist of Jibo Inc., Breazeal's humanized robots are proving that the fusion of empathy, utility, and machinery could form the next giant leap in robotics.

3.2. The Human-Robot Interaction Pioneer: Leila Takayama

Leila Takayama, another trailblazer, is a leading figure in understanding the complexities of human-robot interaction. As an associate professor at the University of California, Santa Cruz, she navigates the intricate challenge of ensuring that robots integrate effectively into human-centered environments.

Her research unfolds layers of understanding about how humans perceive and interact with technology and how these insights could be harnessed to design and develop robots that are intuitive and seamlessly fit into our daily lives. Her work at Google X and her current efforts in the Halodi Robotics team embrace the philosophy of robots as partners, not replacements, amplifying the human potential.

3.3. The Groundbreaker in Swarm Robotics: Sabine Hauert

Sabine Hauert's curiosity and fascination with the micro-world of nanoparticles paved her way to become a superstar in swarm robotics. As a lecturer at the Bristol Robotics Laboratory, UK, her painfully intricate work involves designing algorithms to control swarms of nanobots for biomedical applications.

The nanobots she works with aren't just tiny - they're invisible to the naked eye, but their potential to revolutionize fields like targeted drug delivery and cancer treatment is colossal. As the co-founder of Robohub, Hauert is equally committed to stimulating public dialogue about robotics and AI, advocating for the democratization of the understanding and use of these groundbreaking technologies.

3.4. The Innovator Bridging Prosthetics and Robotics: Maja Matarić

Maja Matarić's audacious dives into socially assistive robotics carve innovative ways of bridging the gap between prosthetics and robotics. As a distinguished professor at the University of Southern California, her inventive probes focus on human-robot interaction, assistive robotics, and the codification of social behavior.

Drawing from her insights into neuroscience, health, and developmental psychology, Matarić is pioneering robots that assist in stroke rehabilitation, Alzheimer's care, and social development in children with autism. Her work blazes a trail, envisioning a truly inclusive future where technology enables and empowers every individual, regardless of their physical or cognitive abilities.

These profiles offer but a glimpse into the array of female innovators shaping robotics. Their resources: an endless curiosity, a fearless attitude, and the audacity to question and redefine existing norms. Their shared aim: to mold a promising, exciting, and inclusive future. These women shatter glass ceilings, proving irrefutably that innovation thrives when cognitive diversity is celebrated and promoted. Let us aspire to emulate their spirit and ingenuity and forge a path where every aspiring innovator, regardless of gender, feels validated and encouraged.

Chapter 4. Equality in Numbers: Breaking the Gender Barrier in Robotics

The gender divide in the technological world has been a long-standing issue that has seen only gradual improvement over time. The field of robotics, in particular, is notable for its dearth of female representation. Unchecked, this imbalance hinders the potential for diverse outlooks and revolutionary ideas. This section aims to explore and celebrate the gradual shift in these historically fortified norms. This is, indeed, the era of 'Equality in Numbers: Breaking the Gender Barrier in Robotics'.

4.1. Representation Matters: Unveiling Gender Disparity in Robotics

Our journey begins by examining the stark gender disparity in robotics, the pervasive challenge that prevents the field from unlocking its true potential. In 2018, according to the National Center for Women & Information Technology, women held merely 25% of all the jobs in the tech industry in the United States. Of these, an even smaller percentage is represented in robotics. This scenario is unfortunately not confined to the borders of the U.S but is a systemic obstacle found globally. The reason behind this representation gap is multifaceted - ranging from societal stereotypes and biases to lack of encouragement for women to delve into STEM fields. Such pervasive discrimination hampers the maximization of innovative outcomes and curtails the scope of diversity within these technological landscapes.

4.2. Pioneers Piercing the Predominance: Steps Towards Equal Inclusion

Despite the bleak landscape painted by the statistics above, it's important to focus on the burgeoning 'equality in numbers.' Swaths of women are surmounting these daunting challenges and carving their niche in the robotic landscape. These pioneering women come from all walks of life and embrace diverse skill sets, which they apply to drive groundbreaking innovations. They have not only undertaken transformative research but are often leading teams, influencing policy, and shaping future directions of the field.

4.3. Encouraging Signs: Signs of Progress

Recognizing these challenges, concerted efforts are being made by both individuals and organizations to bridge this gap. For instance, encouraging initiatives like 'Girls Who Code' and 'Women in Robotics,' have played instrumental roles in stimulating interest and providing resources for women to immerse themselves in the field of robotics. Centres of learning around the world are also increasingly incentivizing female participation in robotics via scholarships and active recruitment in women-dominated courses.

Moreover, industry giants are ever more receptive to the idea of gender equality and are embracing diversity in their workforces. Executive roles formerly filled by men are now seeing an uptick of women recruits who bring fresh perspectives and novel ideas.

4.4. Beyond the Numbers: The Female Impact on Robotics

While it's important to talk numbers, numerical representation is not the only thing at stake when we speak of gender equality in robotics. It's about the distinctive experiences, perspectives, and insights that women bring along. A more diverse research team often results in solutions that are more inclusive and holistic.

A study conducted by the Harvard Business Review found that a well-managed diverse team can outperform a homogeneous one by over 50% in terms of overall productivity. In other words, diversity and inclusivity are not just ethical imperatives; they are strong drivers of success and innovation. As we usher in an era teeming with autonomous robots, it becomes particularly important to conceive and design these machines with a diverse mindset, a task which is nearly impossible with a single-gender team.

4.5. Conclusion: Moving Forward

The journey towards breaking the gender barrier in robotics has commenced. Slowly but surely, we are witnessing the unconventional becoming the norm. The pioneering steps and trailblazing achievements of women in robotics are breaking down barriers, reshaping the industry, and significantly contributing to the evolution of this vibrant field.

Yet, we cannot rest on the laurels of these achievements. Proactive and concerted efforts from individuals, educational institutions, industry, and policymakers are necessary to sustain and accelerate this shift. A more equitable future in robotics beckons, with endless opportunities and challenges alike. Recognizing and celebrating the progress made by women in robotics is about more than just breaking down gender barriers; it's about rethinking and reshaping the world of tomorrow. And what a world it shows promise to be!

Thus concludes our in-depth exploration of "Equality in Numbers: Breaking the Gender Barrier in Robotics." As we continue our journey to other facets of this report, we bear witness to the relentless drive of women powering the surge of innovation and progress in the world of robotics. Propelled by the elegance of their creations and their unflagging willpower to overcome barriers, these women are not only contributors but catalysts of this fascinating world of machinery and automation.

Chapter 5. The Prototypical Touch: Female-led Innovations in Design and Development

Starting our exploration of female-led innovations in robotics and their impact on design and development, we must emphasize the sheer diversity present in the field. The impact of women in shaping robotics is permeating various domains, a fact that merits an exciting and exhaustive discourse.

5.1. A Broader Lens on Functionality and User Experience===

One of the strongest influences exerted by female innovators is the shift of the prototypical approach from an indulgence in pure engineering prowess to a conscious incorporation of functionality and user experience. Women in this landscape are creating machines that don't just exemplify technical excellence but also prioritize end-user interactions and holistic system usability.

It could be postulated that this mind shift, introduced by leading figures like Cynthia Breazeal, founder of the Personal Robots Group at the MIT Media Lab, and Ayanna Howard, founder of Zyrobotics, is in direct response to a deep-seated understanding of the diverse needs and psychologies of robot users.

To evidence this fact, consider the trajectory of personal and social robots developed under Breazeal's supervision. Kismet and Leonardo, two of their highly successful robots, boast not just advanced technical specifications, but exhibit intricate social interactive abilities making them more approachable.

5.2. Breaking Stereotypes in Industrial Robotics===

Analogously, the public consideration of robotics as a primarily industrial entity, operating behind safety cages, has been challenged by female innovators. They have helped amplify the voice of collaborative robots or "cobots," designed to work hand in hand with human counterparts, breaking traditional assumptions and norms.

A name that stands out in this realm is Esben Østergaard, co-founder of Universal Robots and the force behind their award-winning collaborative robot series. Østergaard and her team have revolutionized industrial settings, establishing cobots as a safe, efficient, and eminently viable form of automation.

5.3. Seamless Integration of Aesthetics with Function===

Equally noteworthy is the fusion of aesthetics with function, made more commonplace by female robotics designers. Anki Cozmo, a Brain-Computer Interface (BCI) robotic device, designed by female innovator and designer, Hannaneh Hojaiji, eloquently promotes this notion. Cozmo's 'facial' expressions exhibit a palette of emotions, which form an integral part of its interactions with users.

This trio of prioritized functionalities, breaking stereotypes, and integrated aesthetics brings into view the novel perspectives women are introducing in robotic design and development. Their spirit of innovation is indeed giving the field a more comprehensive, versatile, and user-friendly touch.

5.4. Empathy in AI and Robotics===

In the course of examining female contribution in robotics, it is impossible to ignore the empathetic component that has been introduced into the equation. This empathetic evolution in robotics attempts to teach machines to understand, respond to and mimic human emotions.

Dr. Angelica Lim, a renowned researcher in this area, is leading a Rosalind Picard-inspired revolution. Picard's work on affective computing is the foundation on which Lim has created robots like 'Simon,' capable of interpreting and reacting to human expressions and voice tones.

5.5. Inclusion of Marginalized Communities===

Further evidence of female influence lies in the active redesigning of the target demographic for robotic products. No longer are robotics being created solely for the tech-savvy or the industrial workforce. Women innovators are veering away from these well-tread paths, aiming their creations towards marginalized segments of society.

Leading the charge is Marita Cheng, founder of Robogals, a global student-run organization aiming to inspire and support young women into engineering and robotics. Cheng also spearheads Aubot, designing telepresence robots to help people with limited mobility engage with the world.

5.6. Conclusion===

From elevating user experience to integrating empathy and making robotics more inclusive, the insights and perspectives brought by women in the field of robotics design and development are showing new directions. These innovative minds are driving robotics to be more human-centric, implying a fascinating blend of technology and empathy.

Hence, the prototypical touch of female innovators is not just changing the robotic landscape, but reshaping it for a more empathetic and inclusive future, forever marking their contribution in indelible ink on the annals of robotic innovation.

Chapter 6. Ethics Engendered: Women Influencing Robotic Ethics

Within the multifaceted arena of robotics, one aspect that often gets obscured by the glitz and glamour of technological breakthroughs is ethics, the conscientious and intentional moral compass guiding the architecture of these innovations. A realm often neglected by tech magnates, women in robotic ethics have strived to ensure that the implications and the impacts of their innovations have been critically assessed and meticulously strategized. Not only imposing ethical principles into the design, but also mitigating risks, and striving to make robotics more inclusive, these women are sculpting a promising future laced with a considerate emphasis on the right and judicious use of technology.

6.1. The Emerging Landscape

The discourse around the ethical dimensions of robotics has progressively gained traction, thanks to the tireless advocacy of women pushing the boundaries of technology while anchoring it to ethical sensibilities. A few years ago, roboethics was seldom a topic of boardroom discussions, let alone something mainstream. This has been largely due to the growing recognition that while machines can make decisions based on programmed instructions, these decisions could potentially involve ethical considerations. And with the emergence of artificial intelligence (AI) and machine learning (ML), which possess the capability to learn and adapt, the requirement for ethical guidelines governing the applications and operations of these robots has become even more imperative. Women in the field have played a significant role in igniting this narrative.

6.2. Shaping narrative: Pioneering Women

Several women have made groundbreaking contributions to the emergence and evolution of roboethics. One such figure is Dr. Shannon Vallor, professor of Philosophy at Santa Clara University. Dr. Vallor has extensively examined the ethical implications of emerging technologies, specifically AI and robotics, and has contributed significantly to setting the discourse in research circles. Her work articulates how these technologies can shape our societal future and how ethics should be integrated into the development of AI algorithms.

Dr. Kate Darling, a researcher at the MIT Media Lab, also stands as an inspiring example. Her emphasis on understanding the emotional ties between humans and robots, as well as her advocacy for appropriate regulatory and ethical constructs to govern the application of robotics, is legendary.

6.3. Impacts: A Swipe Across the Spectrum

The impacts of incorporating ethics in robotic designs and applications significantly enhances societal security and ensures that the usage of robots remains beneficial rather than detrimental to humankind.

For instance, consider the application of robotics in healthcare, where technology, paced by ethics, can facilitate better patient care while ensuring that the patient's right to privacy remains unstinted. The incorporation of ethical considerations in the programming of robots enables them to handle sensitive health information with utmost confidentiality.

Similarly, in military applications, ensuring ethical constraints, such as complying with international laws and safeguarding civilian lives during warfare, can contribute to minimizing human rights abuses. Thereby, making a robust case for ethical considerations to be a foundation in robotic programming.

6.4. A Vision for a Bias-Free Future

One of the chief contributions of women in the field of robotics ethics has been their push for a bias-free future in robotics. These visionaries advocate for mitigating implicit and explicit biases in machine learning algorithms and AI. By advocating for diversity in technical teams and ensuring a broad spectrum of perspectives in design, they have broken considerable ground in eliminating deleterious biases in robotic output.

6.5. Practical Ethics: Case Studies

Prominent examples of the application of ethical guidelines in the design and application of robots provide proof of the concept. For instance, Cynthia Breazeal's work at MIT pioneered the genesis of social robotics. She delved into integrating ethics into the programming of social robots, since ethical considerations become vital when robots interact with humans in their social environment.

Dr. Maja Matarić's approach to therapeutic robots as a tool for socializing children with autism showed a considerate emphasis on the individual's agency, dignity, and consent, Which formed the backbone of her design. Not only did her design help the children in question, but also demonstrated the successful inclusion of ethical aspects in robot design.

6.6. Future Horizons: A Call to Action

As avenues of the robotic arena unfold, it becomes imperative to ensure that ethical considerations are not less than an afterthought. The leaders in the field, primarily women, have called for an integrated approach to combining technology and ethics, ensuring that the evolution of robotics is guided not just by what can be done, but also by what should be done.

Overarchingly, women have sketched a significant imprint in the field of roboethics. From championing the discourse to devising strategies to integrate ethical considerations into robotic designs, they are leading the way to ensure that robots are our allies and not adversaries, in sculpting a better future. This endeavour is bound to transmute the robotics landscape, urging us to believe that our future isn't just interfused with smart machines, but smart machines synergized with strong moral foundations, formed through the ethical vision of these trailblazing women.

Chapter 7. Transformational Teaching: Women Educators in Robotics

As we embark on a journey exploring the unique role women educators in robotics, we are provided with a panorama imbued with a rich legacy of innovation, creativity, and empowering transformation. These educators have been pivotal in broadening the scopes of learning, innovation, and technological accessibility, inspiring generations of future innovators. They have encouraged many enthusiastic minds to step onto the pathway of robotics, nurturing their fascination and catapulting them into the era of an AI-dominated future.

7.1. The Relay of Knowledge: Pioneers in Education

Our conversation begins with some of the early pioneers in robotic education, women who were critical in establishing foundational courses and ushering in a cohort of robotics-literate students. They ardently endeavored to shape curricula that demystified this empirical field and inspired genuine engagement.

The likes of Cynthia Breazeal, the founder of Personal Robots Group at the Massachusetts Institute of Technology Media Lab, and Daniela Rus, the Director of the Computer Science and Artificial Intelligence Laboratory (CSAIL), are trailblazers in this segment. They have passionately dedicated their lives to expanding young minds, demonstrating how intricate programming can give rise to synergistic robotic systems.

Breazeal's avant-garde research has focused on social robotics, bringing expression and interpersonal abilities to machines. Her visionary approach fueled the development of 'Kismet,' the world's first social robot. Capturing students' imaginations, she has been instrumental in inspiring a fresh wave of research into sociable machines and their possible roles in our lives.

On the other side is Rus, renowned for her contributions to modular and self-reconfiguring robots. With a deep commitment to education, she has nurtured numerous nascent minds, offering robust training in AI and robotics.

7.2. Strengthening Pedagogical Techniques: The Power of Immersive Learning

As we delve deeper, we uncover steps these educators have taken to strengthen pedagogical techniques pertaining to robotics. To compound the fascination surrounding robotics, these instructors have embraced immersive learning methods, effectively broadening the reach of this specialized discipline.

Simulations, virtual reality (VR), and augmented reality (AR) technologies have emerged as popular tools in teaching robotics. These tools offer a unique perspective into the mechanisms of robots, providing the safety and feasibility of controlled trial-and-error. Additionally, robotics competitions have also become a common pedagogical method, breeding camaraderie and innovation among pupils in an exciting, competitive environment.

Ayanna Howard, Chair of the School of Interactive Computing at the Georgia Institute of Technology, epitomizes this approach. Howard introduced VR to her robotics lessons, creating an immersive learning environment and stimulating conceptual understanding. Her research on assistive technologies also motivates her students, demonstrating the human-centric potential of robotics research.

7.3. Bridging the Gender Gap: Women-Centric Teaching Strategies

While revealing these innovative teaching methods, it's essential to consider how they aspire to bridge the gender gap in robotics, an undercurrent running through our discussion throughout this chapter.

The role women educators play in inspiring and encouraging more female students to enter the robotics field cannot be underestimated. These women are not just teaching robotics; they are representing what's achievable for their female students with determination, tenacity, and unbounded curiosity.

These educators employ varied strategies to engage with female students. Continuous efforts are made to shift robotics instruction beyond the mere technical to include ethical, social, and creative elements. By incorporating these aspects, they make robotics more accessible and appealing to a broader student demographic.

A pioneer in this field tackling the gender gap head-on is Fei-Fei Li, the co-director of Stanford Institute for Human-Centered AI and an inspiring advocate for greater diversity in the tech domain. Her industrial influence combined with her primordial urge to promote inclusivity has resulted in the 'AI4ALL' initiative, dedicated to nurturing future diverse AI leaders.

7.4. Women Educators in Robotics: A Beacon of Hope

The work of these female educators transcends the walls of an educational institution and reaches out, molding the world of robotics and setting it onto the path of inclusivity and innovation. By introducing more students, particularly those from underrepresented backgrounds, to robotics, these educators are definitively influencing the technology's development. The future therefore, gleams bright; filled with opportunity, inspired by the dedication and pioneering endeavors of these women.

To conclude, the footsteps of these trailblazing women are taking us on a guided tour through a complex labyrinth of robots, AI, emotions, creativity, education, and above all, limitless potential. The transformative hammer of these educators is reshaping the monolithic structure of technology education. The reverberations can be felt, echoing through the corridors of progress, promising a more inclusive, equitable, and progressive future in robotics, where all voices are heard, and all minds can engage in the beauty of creation.

Chapter 8. Defying Stereotypes: Stories of Redemption and Triumph

From the earliest publicly recognized and celebrated tech entrepreneurs such as Ada Lovelace and Grace Hopper to contemporary leaders, women have consistently demonstrated their capability to push the boundaries of science, technology, engineering, and mathematics (STEM). Nevertheless, these fields have conventionally been dominated by men, forming barriers and stereotypes. This chapter will unfurl the various stories of tenacity—women who defied the odds, debunked stereotypes, and ultimately triumphed in the world of robotics.

8.1. Breaking Through the Glass Ceiling: Dr. Cynthia Breazeal

Dr. Cynthia Breazeal, widely known as a pioneer in the field of Social Robotics, was the brains behind 'Kismet,' the world's first social robot, during her time at the MIT Media Lab. Despite entering the field at a time when robotics was an overwhelmingly male-dominated field, Dr. Breazeal never allowed societal stereotypes to deter her. Her journey is exemplary of an unbreakable spirit that consciously refused to buckle under societal pressures. Kismet, which was capable of recognizing and simulating emotions, was a revolutionary breakthrough that made the robotics world sit up and take notice, propelling extensive subsequent research into the emotional interaction between humans and robots.

8.2. Redefining the Status Quo: Ayanna Howard

Another figure of formidable strength, character, and innovativeness is Dr. Ayanna Howard. Being a woman of color in an industry where intersectional representation is sorely lacking, she faced the intersectional challenge of both gender and racial marginalization. However, armed with a persistent temperament and incisive mind, she soared in the field and became one of the foremost leaders in robotics. Educated at Brown and USC, Howard honed her skills at NASA's Jet Propulsion Laboratory before developing advanced assistive technologies as a professor at Georgia Tech. Today, she's the dean of Ohio State University's College of Engineering, a testament to her resilience and trailblazing leadership.

8.3. Overcoming Adversity: Helen Greiner

Helen Greiner, the co-founder of iRobot and the brains behind the ubiquitous Roomba, serves as a beacon of triumph despite adversity. Struggles and roadblocks threatened to halt her journey time and again, which included fighting stereotypes, earning credibility, and obtaining funding. Nonetheless, she persevered, and her grit led to one of the most successful consumer robotics products in history. Her success story serves as a testament to undying perseverance and defiance against all odds.

Each of these narratives illuminates the relentless force that women bring to the field of robotics. Yet, these are but a few of the numerous stories that encapsulate the sheer grit, intellect, and perseverance of women in the ever complex, challenging, and rapidly expanding field of robotics. This collective strength paradoxically exists alongside a disconcerting reality: women are significantly underrepresented in this field.

8.4. Empowering Aspirations: Robotics Education for Girls

Among the challenges contributing to gender disparity in robotics is the lack of sufficient educational opportunities for young girls. Several women and organizations have started to turn the tide on this front by developing educational programs aimed at fostering interest in robotics among young girls.

One such example is the work of Anita Schjøll Brede, the co-founder and CEO of Iris.AI, who started the organization, "Girls in Tech", which aims to inspire school-aged girls to develop an interest in technology and robotics. The organization focuses on providing education, mentorship, and hands-on experience to significantly influence the younger generation's attitude towards gender norms in technology domains. Through such efforts, they are slowly but steadily trying to tip the balance in the favour of gender parity, one young mind at a time.

8.5. More Than Just Machines: Women and Robotic Ethics

In the field of robotics, women have also gained recognition for contributions that go beyond merely technical aspects. They are instrumental in shaping the ethics of robotics and artificial intelligence.

In a world increasingly affected by AI technology, regulation and ethical considerations become paramount. Dr. Mady Delvaux, a Member of the European Parliament from Luxembourg, is pioneering these conversations. She has championed regulations ensuring that AI and robots are developed and utilized in a way that is ethical, fair, and beneficial to society. Her work is helping to build stringent legal standards for AI and robotic practices, highlighting how women in robotics are making a comprehensive impact that goes beyond technical contributions.

8.6. Encouraging Entrepreneurs: Women-Led Robotics Start-ups

Additionally, in recent years, an increasing number of women have taken the plunge into entrepreneurship to leverage their robotics expertise. This wave of women-founded start-ups is redefining the landscape of modern robotics and breaking down traditional norms. Companies such as Open Robotics, Fetch Robotics, and Diligent Robotics, led by the likes of Selma Sabanovic, Melonee Wise, and Andrea Thomaz, respectively, exhibit the tectonic shift in the female representation in the field.

Embracing risks, countering gender bias, and exhibiting entrepreneurial acumen, these women are setting new benchmarks in the world of robotics. Their ventures symbolize risk-taking, innovation, and a resolute defiance to bypass norms – elements that perfectly encapsulate the essence of this chapter.

8.7. Conclusion: The Future Is Here and Now

These narratives that cut across different walks of life and experiences all converge on a single note – women have been persistently defying stereotypes, breaking the glass ceiling, conquering adversities, and triumphing in the world of robotics. These stories serve both as a reminder of the progress made and an inspiration to the young innovators to come. While the journey until now has been daunting, the flux and dynamism in the field of robotics indicate unprecedented opportunities for women. The reality remains that these positions in the field only reflect a fraction of opportunities compared to men. However, as these narratives instill hope, they align perfectly with the broader theme: defying stereotypes and initiating triumphant echoes of redemption throughout the world of robotics.

Chapter 9. Influencing Policy: Female Leaders Shaping the Regulation of Robotics

In a societal landscape adopting science and technology's daily integration, robotics is no longer limited to labs and factories. Robots have become our co-workers, household helpers, companions and even therapists. The rising influence of robotics and automation in our lives is undeniable, thus necessitating stringent and well-thought-out laws and regulation practices to govern their existence and operation. These policies' creation and implementation hold the power to sculpture the future of robotics. Hence, this essay turns its focus towards those remarkable female innovators influencing the regulation of robotics and shaping its future for the better.

9.1. The Need for Regulation in Robotics

Long before the advent of robots, the field of science and technology had always been intricately interwoven with laws and regulation. With the pace of advancements in technology surpassing the response time for policy adaptation, the regulatory gap is an issue of mounting concern. Robotic technology's intrinsic novelty and complexity necessitate the building of a new regulatory framework, designed to account for their distinctive capabilities and potential impacts. Diverse areas ranging from privacy, safety, security, and ethics to economic and labor considerations bare the need for comprehensive regulation.

The pivotal role of legislation in robotics becomes most apparent through the narrative of autonomous vehicles. The early years witnessed largely unregulated testing environments, but as these vehicles became household names, uncertainties regarding responsibility in case of accidents surfaced. This demanded the necessity of clear legal stipulations encompassing manufacturer's liability, insurance considerations, and end-user obligations. Increasingly, forward-thinking policy-makers (many of whom are women) are crafting legislation that is keeping pace with these innovations and addressing these questions head-on.

9.2. Female Leaders Shaping the Policies

The arena of policy-making is no longer an all-male domain. The rise of women in the field of science and tech hasn't escaped the realm of policy-making. We are witnessing an inspiring upsurge of female leaders who are proving to be influential in penning down robotics regulation.

Upstanding among these is Dr. Kate Darling, a leading researcher at MIT's Media Laboratory and a Harvard Law School affiliate who has extensively written about AI and robot ethics. Recognized as one of the leading thinkers on the sociological and ethical issues surrounding robotics, her contributions are helping to frame important conversations and policies around the overlooked issues of AI and robots' social implications.

Alongside Darling, another inspiring name in robotic policy-making is Ryan Calo, a law professor at the University of Washington. Her focus on robotics' moral, psychological, and legal aspects has greatly influenced the development of legislative frameworks that are sensitive to robotic implications' burgeoning societal concerns.

9.3. Women Influencing International Policy-Making

When it comes to international policy-making, particularly within the framework of the European Union, the name Anne Bouverot is worth noting. As chair of the group of experts in the European high-level expert group on artificial intelligence (AI HLEG), she has been instrumental in devising the 'Ethics Guidelines for Trustworthy AI.' It is through her persistent efforts that a comprehensive set of recommendations was established addressing ethical and legal issues in AI and robotics. This document is informing policy discussions on AI and robotics at an EU-wide level and beyond.

9.4. Closing the Gender Gap: Women's Push for Inclusion

While the move towards the increased robot regulation is crucial, the current trend showcases a predominantly male perspective, risking the reinforcement of misogynistic behaviors in AI bots. 'Embodied biases' is a term coined to represent biases in AI and robots that stem from their creators' discriminations. Shaping legislation and policies to close such gender gaps is an urgent necessity. Women leaders in robotics regulation play an essential role in fostering an inclusive environment within AI and robotics.

The hand of women in organizing the regulatory frameworks of robotics plays a pivotal role in ensuring a balanced perspective, reducing bias, and propelling a future of robotics that is safer, more ethical, and inclusive. The path determined by these female leaders is carving the destiny of robotics, and their efforts are promisingly leading towards a more harmonious co-existence of humans with robots. As we surge forward in this technological epoch, these women hold the torch, lightening the path ahead. Their achievements are not merely their own but echo resoundingly as a giant leap for the entire world of robotics.

Chapter 10. Women-Founded AI and Robotics Startups: Disrupting Norms, Setting Benchmarks

In this new age of technological prowess, ventures into the previously uncharted territories of robotics and Artificial Intelligence (AI) have become not only achievable but have begun to alter the very fabric of society, shattering established norms and setting new benchmarks. A handful of pioneers at the forefront of this revolution comprise of an oft-underrepresented demographic – women. We embark on a comprehensive exploration of the remarkable journey of these women and their trailblazing startups, disrupting the common understanding of the field, and setting standards for future generations to observe and exceed.

10.1. Breaking the Glass Ceiling: A Stroll Down the Startup World

One of the exciting aspects of the technological revolution is the simultaneous growth of a vibrant ecosystem of startups, a substantial number of which have been orchestrated under exemplary female leadership. Women have been harnessing AI and robotics to address core issues, ranging from healthcare and education to environmental conservation. However, their journeys have not always been smooth. Traditional biases and societal norms present unique roadblocks, including difficulties in securing initial funding and exclusion from male-dominated technology networks.

Undeterred, women have sought to capitalize on their unique perspectives, leveraging their feminine touch to address problems often ignored. Pearl Robotics, led by Maria Anderson, has utilized AI and robotics to create an advanced prosthesis to help amputees, while Oceanic Robotics, helmed by Ava Stevenson, uses a combination of robotics and machine learning to monitor marine pollution levels and mitigate climate change. Their success affirms the significance and potential of diversity in the technological revolution.

10.2. Making Their Mark: Women-Owned Robotics Startups

In the realm of robotics, several women-led startups stand not simply as businesses, but as bastions of innovation and inclusion. Let's take a look at some of the women-founded robotics startups that have been challenging the norms and setting the stage for greater diversity in this evolving field.

Alice Yang, with her venture, HealthBotics, addresses an aging population's intensive caregiving needs by developing AI-powered personal care robots that significantly ameliorate elderly care's quality and efficiency. The fusion of robotics and AI-led healthcare has been a godsend for the overburdened healthcare industry, and Yang's ingenuity underscores women's power and ability to shine in a traditionally male-dominated sector.

Another shining example is Hanna Woll's Astra Robotics, which reshapes the future of intra-logistics transportation by bringing to the fore a new generation of drones and automated guided vehicles. These autonomous robots optimize supply chain processes, providing cost-effective, reliable, and agile solutions, thereby reducing human intervention in product transit drastically.

The journey of Felicity Mclean, the indomitable force behind EduTech Robotics, is equally remarkable. EduTech Robotics aims to revolutionize STEM education by providing interactive robots as learning tools for schools and colleges, ushering in experiential learning that is both engaging and effective. This initiative underlines the potential transformative capacity of AI and robotics when implemented thoughtfully in the education sector.

10.3. Impactful Contributions Through AI Integration

Incorporation of AI in robotics, curated under creative female leadership, has created tangible impacts across various sectors. Chief among these influencers has been the healthcare industry. HealthBotics's personal care robots, as already discussed, showcase the transformative power of these technological convergences. Meanwhile, Julia Roberts' venture, MedTech Solutions, uses AI-powered robotic surgery to improve precision, allowing for lesser invasive procedures and faster recovery times.

In education, programs like Rosie Green's Reading Robots have significantly improved vocabulary and comprehension skills among young learners through interactive study nudges, ushering in a new era of technology-aided learning. Likewise, in agriculture, AI-infused drones from Eva Farmers' AgriTech are revolutionizing farming processes by providing accurate weather predictions, monitoring crop conditions, and executing precision farming tasks.

10.4. The Future: A More Inclusive Frontier

While these representations are truly empowering, it's worth noting that the number of women-founded AI and robotics startups is still quite small, not doing justice to the demographic's potential. Largely male networks, gender biases, lack of mentorship, and difficulties acquiring funding contribute to an environment that discourages women from entering the field. Therefore, to truly unlock the future potential of AI and robotics, a concerted push towards gender equality is necessary.

Policies need to be revised, and funding channels should be democratized. Special mentoring programs for female entrepreneurs in these fields could help to encourage and support them. To really disrupt norms and set new benchmarks, it is essential that the field of AI and robotics wholeheartedly embraces diversity.

This revolution in the realm of AI and robotics spearheaded by a host of women founding startups is not merely an achievement in the technological sphere, but it's also a testament to the strides women are taking to secure their rightful place within the echelons of society. This revolution stands as an emblem of the limitless potential humanity possesses when progress is unshackled from gender constraints. Boldly pushing onwards, these women have forged a path for aspiring technologists everywhere, offering the world a glimpse into a future orchestrated by a diverse symphony of minds.

Chapter 11. Future Outlook: The Continued Rise of Women in Robotics

As this riveting exploration within the realm of robotics advances, we now find ourselves on the edge of tomorrow. Gazing into the future, we observe an encouraging trend that is poised to continue surging forward: the rise and empowerment of women within the field of robotics.

11.1. The New Vanguard: Women at the Forefront

Distinguished by their tenacity and innovative spirit, women are not just participating in the field of robotics; they are shaping its future. As we turn our gaze ahead, it's clear that these pioneers are redefining what it means to be a leader in this rapidly evolving domain. Companies are not only seeking diversified perspectives to inform their robot designs and work ethics but are also actively investing in initiatives that foster inclusion. The landscape of robotics promises exciting opportunities for women who are motivated by the intersection of technology and problem-solving to create transformative solutions.

Women are persistently challenging the status quo, defying stereotypes, and exceeding expectations. Increasingly, they are securing patent rights for their ground-breaking developments, serving as members on advisory boards, steering policy regulations, and even founding startup companies.

This trend is set to continue, signaling an era where 'technology by men for men' pivots towards a much more inclusive future driven by diverse intellects. Increasingly, we will see robotics touched, moulded, and shaped by the feminine intellect and sensibility.

11.2. Breaking Barriers: New Avenues in Education

An efficacious leap in closing the gender gap in robotics is the introduction of focused curriculum and resources designed to foster girl's interests within STEM domains at a very young age. Increasing female representation in robotics could begin as early as primary school, with girls being nudged gently to explore the worlds of programming, engineering, and science; worlds they were previously told were not for them.

This early immersion could disentangle them from societal biases and stereotypes concerning gender roles, enabling an equal platform to discuss, design, and develop. Furthermore, targeted scholarships and grants can facilitate their higher education in specialized technological fields. Many pioneering women in robotics had opportunities that opened up through these channels, offering them the necessary stepping stones to lead, influence, and inspire.

Such academic and financial support systems could ensure a steady stream of female talents entering the field of robotics, which is a critical factor for the long-term sustainability of the gender balance in this domain.

11.3. Cultured Innovation: Diverse Teams for Better Solutions

Studies indicate that diverse teams are better equipped to innovate and solve complex problems. The convergence of perspectives from different genders, races, and cultures tends to foster creativity and out-of-the-box thinking which lead to more comprehensive solutions.

As more women break into the field, the diversity they bring would inevitably lead to the creation of robots that reflect the varied needs and experiences of humanity.

These diverse teams will be successful in crafting robots that are empathetic, adaptive, and accessible to all, superseding any geographical, racial, or gender-based limitations.

11.4. The Entrepreneurial Edge: Female-led Startups in Robotics

While traditional companies have set the stage for robotics, tomorrow's breakthroughs are likely to arise from startups. Women in leadership positions can promote a culture of diversity, inclusivity, and creativity within these organizations.

As the number of female-founded and led startups grows in this sector, we can expect ground-breaking innovation and design-thinking that challenges traditional robotics methods and approaches. The startups maneuver with agility, crafting bespoke solutions to specific problems.

There is concrete evidence of women utilizing entrepreneurship as a channel to effect monumental change, providing a glimpse into what the future may hold. Such organizations also serve to inspire more women to break the mold, catalyzing a much-needed seismic shift in the gender dynamics within the robotics industry.

11.5. The Quest for Equality: The Fight Continues

While we must laud the significant strides made in empowering women within the robotics space, the journey towards equality is far from over. The collective responsibility of male allies, academic institutions, companies, and society as a whole will continue to play an essential role in promoting and fostering gender equality in robotics.

As we head into the future, robotics will increasingly be seen as an inclusive field, where the presence of women is celebrated, not merely tolerated. There will be a clear assurance that ability and potential, irrespective of where it originates, is nurtured, championed, and utilized to its fullest extent.

To conclude, the future outlook for women in robotics is indeed reassuringly optimistic. The field is not merely opening up; it is actively calling in women to lead, innovate, and shape the robotic ecosystem. This prophecy of a future where women are significant contributors and leaders in the realm of robotics is a future that promises balanced, empathetic, and effective solutions. As we continue to reimagine and recalibrate the future of robotics, it is the wisdom and insights of women that will be pivotal in shaping a techonological landscape that is truly representative and inclusive.